Space Machines

Library of Congress Number: 78-26991

    2  3  4  5  6  7  8  9  0  83  82  81  80  79

Printed in the United States of America.

**Library of Congress Cataloging in Publication Data**

Ciupik, Larry A.
    Space machines.

    SUMMARY: Describes a variety of equipment for
use in space, including space laboratories and
stations, mining machines, and transporters.
    1.  Astronautics — Equipment and supplies —
Juvenile literature.  2.  Space colonies — Equipment
and supplies — Juvenile literature.  [1.  Astro-
nautics — Equipment and supplies]  I.  Seevers,
James A., joint author.  II.  Title.
TL793.C56   629.44   78-26991
ISBN 0-8172-1325-2

Cover illustration: Jerry Scott

Photographs/Artwork appear through the courtesy of the
    following:

Aérospatiale, Division Systèmes Balistiques et Spatiaux:
    pp. 3, 6 (bottom)
Astronomy Magazine artwork (p. 29) by
    Thomas Schroeder, copyright © 1977
    by AstroMedia Corp. All rights reserved.
Lockheed Missiles and Space Company: p. 23
National Aeronautics and Space Administration: pp. 4, 5, 6 (top),
    7, 8, 9, 10, 11, 12, 13, 14, 15, 16, 17, 24, 26, 27, 28
Rockwell International, Space Division: pp. 18, 19, 20, 21, 22, 25

# space
# MACHINES

Larry A. Ciupik, M.S., Astronomer
James A. Seevers, M.S., Astronomer

RAINTREE CHILDRENS BOOKS
Milwaukee • Toronto • Melbourne • London

For many years space travel was just a dream. But now many space machines spin high above the earth. Some of them fly to other planets. Some space machines have taken people to the moon.

One kind of space machine is called a satellite. Some satellites were like huge aluminum foil balloons. They bounced radio programs from one place to another on the ground. Now we can beam radio and television programs from space to the whole world.

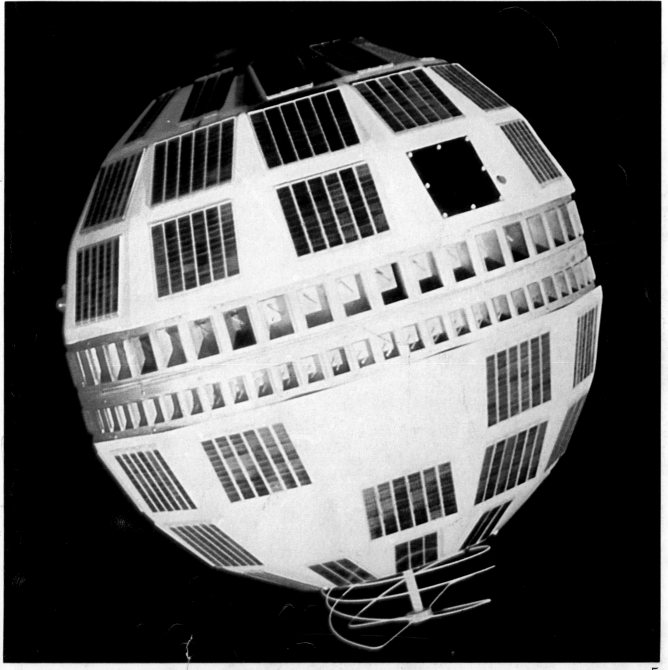

Weather satellites fly high above the earth. They take pictures of the earth. Some take pictures of half the earth at once. Some watch only part of the United States. You can see weather satellite pictures on the news on television.

One satellite is named Landsat. It means "land satellite." It takes pictures of the earth in different colors. Each color tells something. Red shows a healthy forest. Landsat can help tell if crops are healthy. It can also help find places to drill for oil.

A few space machines have been to the moon. The first satellite went to the moon to take pictures. It took pictures and then crashed on the moon. Next, machines landed on the moon to take pictures and dig up moon rocks. Other machines went around and around the moon to take pictures.

After many years of satellite tests, people went into space. One early spaceship was called Mercury. It carried one person. Next, two people went around the earth in a spaceship. It was called Gemini. Sometimes one astronaut would get out of the ship. Astronauts were tied to the ship by a long cord.

Next, a spaceship that could carry three astronauts was built. It was called Apollo. It went to the moon. It took all three astronauts to fly the Apollo. One stayed in orbit around the moon. The other two landed on the moon. A car called the Rover was used on the last three Apollo trips to the moon.

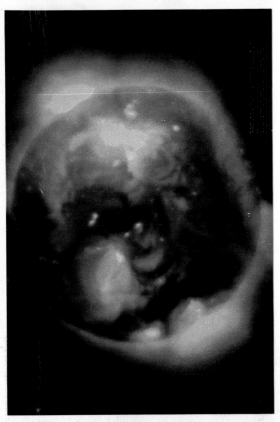

Some space machines watch the sun. They check the sun for explosions on its surface. They also measure the heat of the sun. Other machines fly between the planets. These machines can watch for X rays that could hurt someone in a spaceship.

One spaceship flew past the planet Venus. When it passed Venus, it took pictures of the clouds. The spaceship then flew to the planet Mercury. As it flew past Mercury, it took pictures of craters and mountains.

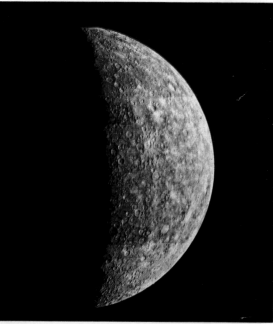

Two spaceships landed on the planet Mars. They are called Viking. Each Viking spaceship had two parts. One part stayed in orbit above Mars. The other part landed on the ground of Mars. The part that landed is called a lander.

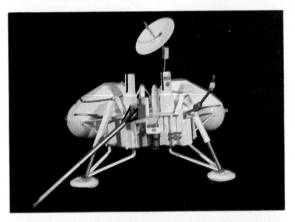

Each Viking lander took pictures of the ground. The landers also tested the soil to see if there was anything living on Mars. The spaceship that stayed in orbit around Mars took pictures of the ground.

Two spaceships flew past the planet Jupiter. They are named Pioneer. The Pioneer ships took pictures of Jupiter and four of its moons. They then flew to the planet Saturn.

Two other spaceships were to fly past Saturn also. They are named Voyager.

The Pioneer spaceship has a picture message on it. The Voyager spaceships carry a small record with earth sounds and messages.

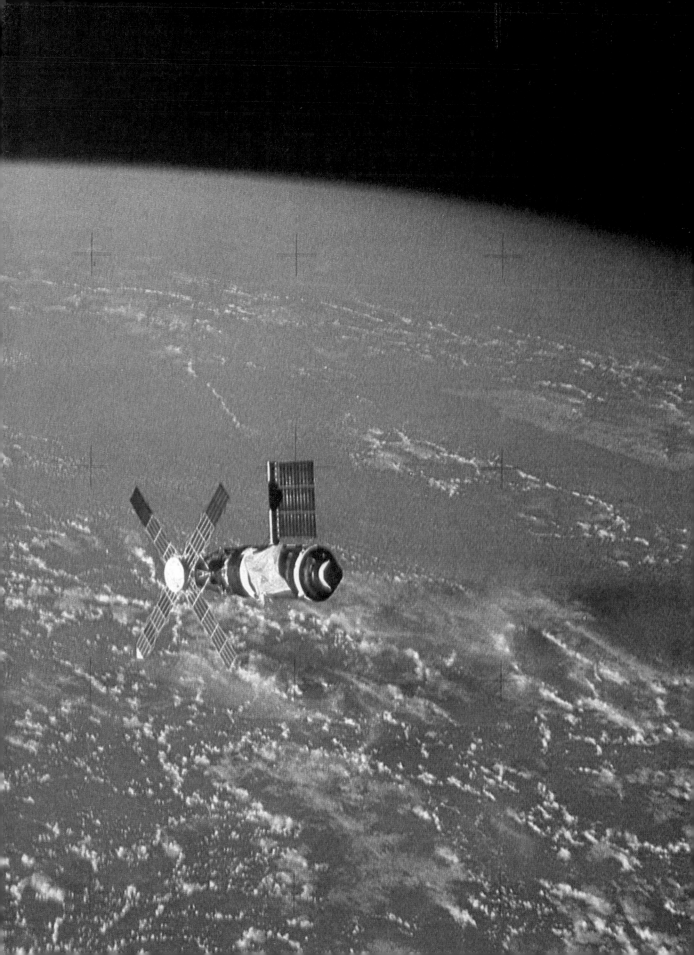

Machines cannot do everything well. People are needed to work in space. Astronauts who want to stay in space for a long time need a place to live. A space station is like a house in space. The first American space station was called Skylab. It had a kitchen, a bedroom, a bathroom, and a workshop. Three Skylab astronauts stayed in space for 84 days. They took pictures of the earth, sun, and stars.

The Soviet Union and the United States made a space station called Apollo-Soyuz. Five astronauts worked together on science projects. The Soviets also have a space station. Soviet astronauts stayed in space for 140 days.

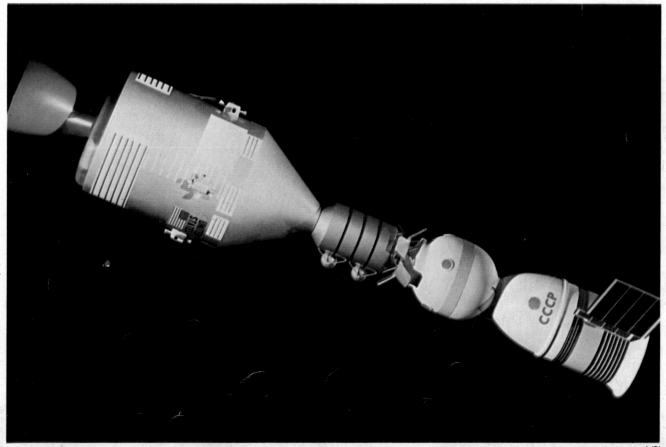

It is easier to put satellites into space using the Space Shuttle. It is like a space truck that can be used again and again. It can carry 32 tons of cargo into space. Several satellites can go up into space on one flight. The cargo area is 15 feet (5 meters) wide and 60 feet (18 meters) long.

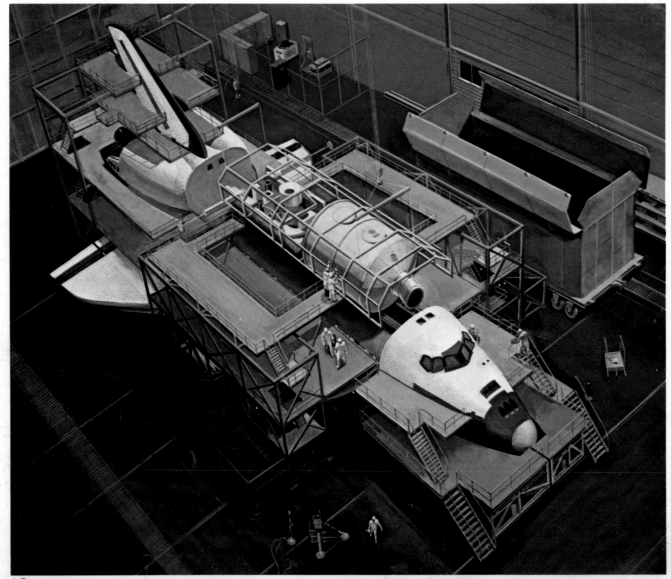

The Space Shuttle orbiter looks something like an airplane. It will have three pilots, and four other people can ride in it. When it comes down from orbit, it lands like an airplane. The Space Shuttle was tested on top of a large airplane.

The Space Shuttle is launched like a rocket. Many large machines are needed to get the shuttle ready to be launched. The large tank holds fuel for the three rocket engines in the orbiter. Two rockets on the side help push it off the ground.

The Space Shuttle will have many uses. It can put new satellites into space. It can return or repair old satellites. Someday it will be used like an airline. The shuttle will make it cost less to send satellites into space.

Many scientists want to do experiments in space. They will be able to work in Spacelab. It is like Skylab, but it is carried by a shuttle. At first, it will land with the shuttle. But it could be made into a big space station. Eleven countries in Europe and the United States are all helping to build Spacelab.

People who study the stars are called astronomers. Some astronomers are building a large telescope to go into space. The space telescope will see farther than any telescope on earth.

Many scientists want to go to the moon. They would explore the moon and build special telescopes there. Some people want to mine the moon for rocks that have aluminum, iron, oxygen, and glass. Large machines will dig up the moon rocks. Another machine will send the rocks from the moon to a space factory. The space factory will have a solar furnace to melt the moon rocks.

At the space factory, workers will make metal parts for large sun satellites. These satellites will be miles long. They will collect sunlight and make it into electricity. The solar power satellite will send energy to the earth.

Space workers building power satellites will make a space city to live in. In will be nicer than their factory space station. They will bring families from earth to live there. The first space colony will be about one mile across. Ten thousand people could live inside the colony.

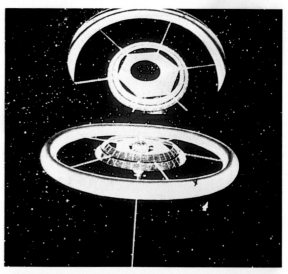

Future space colonies will be
made in many shapes and sizes.
Some of them could be 75 miles
(121 kilometers) long. All space
colonies will have air, water, and
sunlight. The colonies will have
lakes and streams. Trees and
flowers will grow there. Farmers
will grow crops and raise animals
on space farms.

After people live in space colonies for a while, they may want to visit the planet Mars. They may even make a Mars colony. Mars does not have enough air to breathe. They would need a space suit to walk outside if they lived on Mars.

In science fiction stories, it seems easy to fly to stars. But the stars are very far away. The closest star to the sun is billions of miles away. The fastest rocket would take over a million years to get there.

# GLOSSARY

aluminum — A light metal used to build things.

Apollo — A space project that carried people to the moon.

Apollo-Soyuz — A space station that was built by the Soviet Union and the United States.

astronomer — A person who studies the universe.

crater — A large pit on a planet or moon.

Gemini — A spaceship that carried two people.

Landsat — "Land satellite." Landsat takes pictures of the earth in different colors.

launch — To send up into space.

Mercury — An early spaceship that carried people into space.

orbit — To move around and around something. Also, the path a satellite takes around something.

Pioneer — A spaceship that took pictures of the planet Jupiter.

planet — One of the nine large objects that circle the sun. The earth is a planet.

satellite — An object made by people that moves in orbit around another object.

Skylab — The first American space station. Three astronauts stayed in Skylab for 84 days.

solar — Of the sun.

Spacelab — A large space station that is being built by the United States and eleven countries in Europe.

Space Shuttle — A spaceship that looks like an airplane. The Space Shuttle can go from place to place in space. Someday the Space Shuttle may be used like an airline.

space telescope — A large telescope that will go into space.

telescope — An instrument used to make objects in the sky look bigger and closer.

Viking — The name of spaceships that took pictures of and landed on Mars.

Voyager — Spaceships that were to fly past the planet Saturn. They carry a small record with earth sounds and messages.

weather satellite — Satellites that fly high above the earth and take pictures. The pictures help us tell what the weather is like.

# INDEX